I bring the mail.

Mail may come in early.

An airplane brought this mail.

I have something with Mr. Mayfair's name on it.

I ring the bell.

I get no answer.

I'll gladly save Mr. Mayfair's mail.

This job may seem endless.
But do I like it?

Yes, thank goodness!